STRONG KIDS HEALTHY PLATE

Delicious Dairy

Katie Marsico

CHERRY LAKE PRESS

Published in the United States of America
by Cherry Lake Publishing Group
Ann Arbor, Michigan
www.cherrylakepublishing.com

Content Adviser: Debbie Fetter, PhD, Assistant Professor of Teaching Nutrition, University of California, Davis
Reading Adviser: Marla Conn, MS, Ed, Literacy specialist, Read-Ability, Inc.

Photo Credits: ©Africa Studio/Shutterstock.com, front cover; ©MaraZe/Shutterstock.com, 1; ©Lapina/Shutterstock.com, 4; ©GOLFX/Shutterstock.com, 6; ©pixelheadphoto digitalskillet/Shutterstock.com, 8; ©USDA/ChooseMyPlate.gov, 10; ©karelnoppe/Shutterstock.com, 12; ©David Tonelson/Shutterstock.com, 14; ©dcwcreations/Shutterstock.com, 16; ©Madhourse/Shutterstock.com, 18; ©Iakov Filimonov/Shutterstock.com, 20

Cherry Lake Press is an imprint of Cherry Lake Publishing Group.

Library of Congress Cataloging-in-Publication Data

Names: Marsico, Katie, 1980- author.
Title: Delicious dairy / Katie Marsico.
Description: Ann Arbor, Michigan : Cherry Lake Publishing, 2020. | Series: Strong kids healthy plate | Includes index. | Audience: Grades K-1 | Summary: "Dairy products keep you strong and healthy. Discover the variety of tasty dairy products you can eat to maintain a healthy diet. Content encourages balance and making healthy choices. This level 3 guided reader is based on the U.S. government's diet recommendations. Readers will develop word recognition and reading skills while learning about food and where it comes from. Includes table of contents, glossary, index, author biographies, and word list for home and school connection"—Provided by publisher.
Identifiers: LCCN 2020002744 (print) | LCCN 2020002745 (ebook) | ISBN 9781534168626 (hardcover) | ISBN 9781534170308 (paperback) | ISBN 9781534172142 (pdf) | ISBN 9781534173989 (ebook)
Subjects: LCSH: Dairy products in human nutrition—Juvenile literature. | Nutrition—Juvenile literature. | Children—Nutrition—Requirements—Juvenile literature.
Classification: LCC QP144.M54 M37 2020 (print) | LCC QP144.M54 (ebook) | DDC 613.2/77—dc23
LC record available at https://lccn.loc.gov/2020002744
LC ebook record available at https://lccn.loc.gov/2020002745

Cherry Lake Publishing Group would like to acknowledge the work of the Partnership for 21st Century Learning, a Network of Battelle for Kids. Please visit http://www.battelleforkids.org/networks/p21 for more information.

Printed in the United States of America
Corporate Graphics

Table of Contents

About the Author

Katie Marsico is an author of nonfiction books for children and young adults. She lives outside of Chicago, Illinois, with her husband and children.

What type of cheese is this?

What Is Dairy?

Do you like yogurt and cheese? Do you drink milk?

These are **dairy** foods. Dairy foods are milk or are made from milk.

Some dairy farmers get milk from cows.

Goats and sheep are also dairy animals.

Cream cheese is dairy. So are butter and pudding. Many people enjoy ice cream! What are some dairy treats?

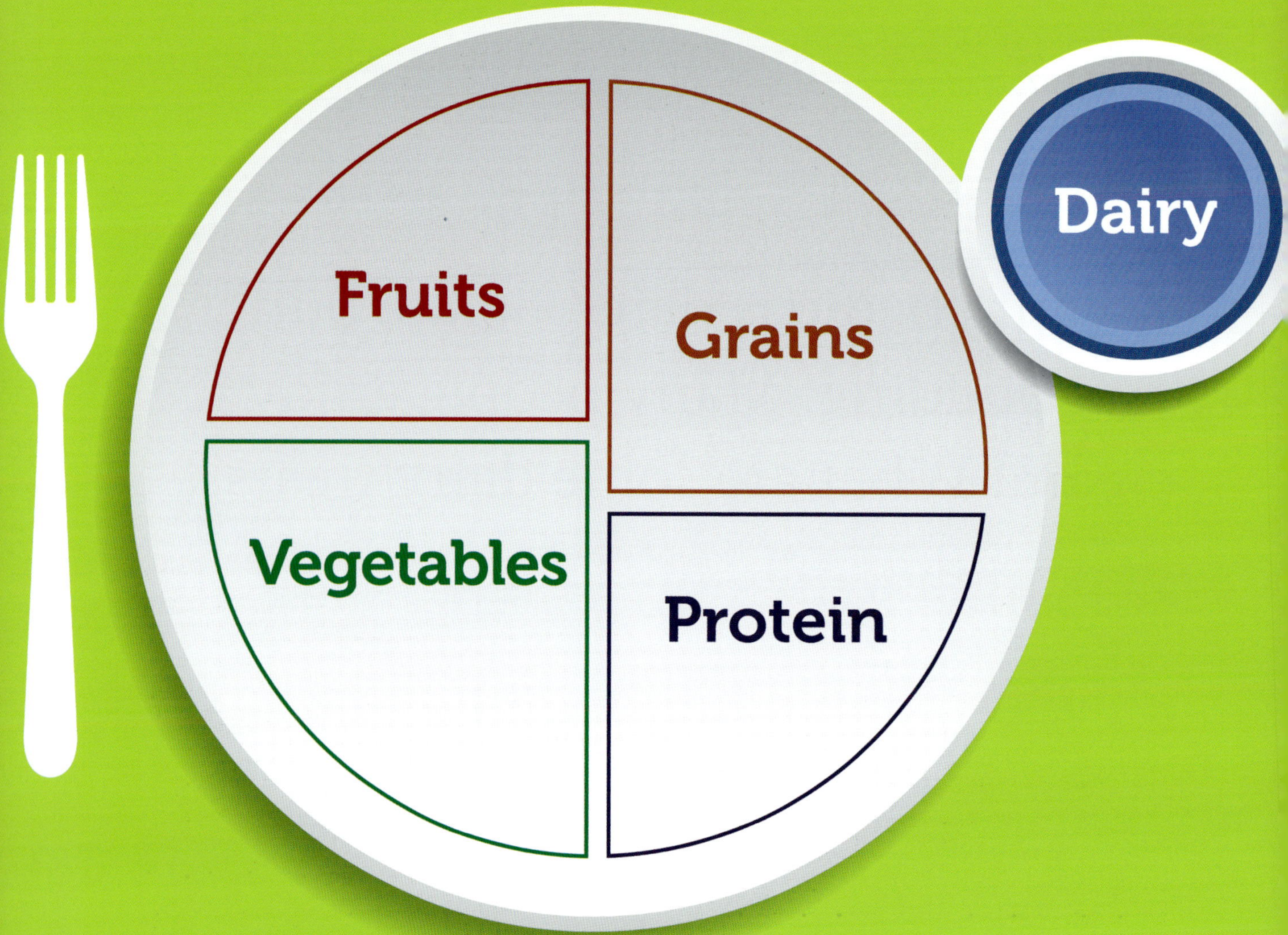

ChooseMyPlate.gov

An Important Part of Your Plate

Dairy is one of five main **food groups**. You need all five to be healthy.

The food groups make up a **balanced diet**. A balanced diet helps you grow and stay strong.

Name your favorite dairy food.

How Does Dairy Help?

Dairy gives you **calcium**. Calcium keeps your teeth healthy. It protects your bones.

HORIZON CA
4.79
25 SEP 2019 22:59 06176 EH3 5L SELL BY
06 OCT 2019 15:17 06176 EH3 6L SELL BY
21 SEP 2019 16:11 06176 EH3 6L
HIGH VITAMIN D
SUPPORTS STRONG BONES
HORIZON
ORGANIC
1% MILKFAT
ORGANIC
PASTURE-RAISED
LOWFAT
MILK
2% MILKFAT
HIGH VITAMIN D
ORGANIC
PASTURE-RAISED
REDUCED FAT
MILK
HORIZON ORGANIC MEANS
NO TOXIC PESTICIDES
PASTURE-RAISED COWS
HALF GALLON

Vitamin D is added to some dairy. Vitamin D also builds strong bones.

Potassium is in dairy too. Potassium supports heart health.

Does milk build strong bones?

100% Farmer Owned
All Proceeds Go To Dairy Farmers and Their Families
KEMPS
QUALITY SINCE 1914
2% MILKFAT MIN.
2% Lowfat Cottage Cheese
EXCELLENT SOURCE of PROTEIN
13g per serving
GRADE A PASTEURIZED
NET WT 22 OZ (624g)
SMALL CURD • NATURALLY CULTURED

Dig into a Healthy Diet!

How much dairy do you need? Try to have 2 1/2 to 3 cups (.6 to .7 liter) each day. Low-fat or fat-free foods are best.

What counts as a serving of dairy? A cup of milk or frozen yogurt does. So does 1/3 cup (79 milliliters) of **shredded** cheese.

Dairy is an important part of your diet. It's a tasty one too!

What dairy foods will you eat today? How will you make healthy choices?

Glossary

balanced diet (BAL-uhnsd DYE-it) eating just the right amounts of different foods

calcium (KAL-see-uhm) a substance found in dairy that helps build healthy teeth and bones

dairy (DAIR-ee) food made mainly of or from milk

food groups (FOOD GROOPS) groups of different foods that people should have in their diet

potassium (puh-TAH-see-uhm) a substance found in dairy products that supports heart health

shredded (SHRED-uhd) cut into long, thin pieces

vitamin D (VYE-tuh-min DEE) a substance found in dairy that helps bones grow and stay strong

Home and School Connection

a
added
all
also
an
and
animals
are
as
balanced
be
best
bones
build
butter
calcium
cheese
choices
counts
cows
cream
cup
cups
dairy
day
diet
dig
do
does
drink
each
eat
enjoy
farmers
fat
favorite
five
food
free
from
frozen
get
gives
goats
groups
grow
have
health
healthy
heart
help
how
ice
important
in
into
is
it
it's
keeps
like
low
made
main
make
many
milk
much
need
of
one
or
part
people
plate
potassium
protects
pudding
serving
sheep
shredded
so
some
stay
strong
supports
tasty
teeth
the
these
this
to
today
too
treats
try
type
up
vitamin
what
will
yogurt
you
your

Find Out More

Book

McMullen, Gemma. *Milk and Dairy Products*. New York, NY: AV2 by Weigl, 2017.

Website

USDA—MyPlate Kids' Place

www.choosemyplate.gov/browse-by-audience/view-all-audiences/children/kids

Use games and online learning tools to find out more about dairy and the other food groups.

Index